Developing
Essential Understanding
of
Multiplication and Division
for Teaching Mathematics *in*
Grades 3–5

Albert Dean Otto
Illinois State University
Normal, Illinois

Janet H. Caldwell
Rowan University
Glassboro, New Jersey

Edward C. Rathmell
Volume Editor
University of Northern Iowa
Cedar Falls, Iowa

Cheryl Ann Lubinski
Illinois State University
Normal, Illinois

Rose Mary Zbiek
Series Editor
The Pennsylvania State University
University Park, Pennsylvania

Sarah Wallus Hancock
Maplewood Richmond Heights
School District
Saint Louis, Missouri

NATIONAL COUNCIL OF
TEACHERS OF MATHEMATICS

Copyright © 2011 by
The National Council of Teachers of Mathematics, Inc.
1906 Association Drive, Reston, VA 20191-1502
(703) 620-9840; (800) 235-7566; www.nctm.org
All rights reserved

Library of Congress Cataloging-in-Publication Data

Developing essential understanding of multiplication and division for
teaching mathematics in grades 3-5 / Albert Dean Otto ... [et al.].
    p. cm.
 Includes bibliographical references.
 ISBN 978-0-87353-667-7
 1. Multiplication—Study and teaching (Elementary) 2. Division—Study
and teaching (Elementary) I. Otto, Albert Dean. II. National Council of
Teachers of Mathematics.
 QA115.D48 2011
 372.7'049--dc22
                                        2010053665

The National Council of Teachers of Mathematics is a public voice of mathematics
education, supporting teachers to ensure equitable mathematics learning of the
highest quality for all students through vision, leadership, professional development,
and research.

Printed in the United States of America

Contents

Foreword

Teaching mathematics in prekindergarten–grade 12 requires a special understanding of mathematics. Effective teachers of mathematics think about and beyond the content that they teach, seeking explanations and making connections to other topics, both inside and outside mathematics. Students meet curriculum and achievement expectations when they work with teachers who know what mathematics is important for each topic that they teach.

The National Council of Teachers of Mathematics (NCTM) presents the Essential Understanding Series in tandem with a call to focus the school mathematics curriculum in the spirit of *Curriculum Focal Points for Prekindergarten through Grade 8 Mathematics: A Quest for Coherence*, published in 2006, and *Focus in High School Mathematics: Reasoning and Sense Making,* released in 2009. The Essential Understanding books are a resource for individual teachers and groups of colleagues interested in engaging in mathematical thinking to enrich and extend their own knowledge of particular mathematics topics in ways that benefit their work with students. The topic of each book is an area of mathematics that is difficult for students to learn, challenging to teach, and critical for students' success as learners and in their future lives and careers.

Drawing on their experiences as teachers, researchers, and mathematicians, the authors have identified the big ideas that are at the heart of each book's topic. A set of essential understandings— mathematical points that capture the essence of the topic—fleshes out each big idea. Taken collectively, the big ideas and essential understandings give a view of a mathematics that is focused, connected, and useful to teachers. Links to topics that students encounter earlier and later in school mathematics and to instruction and assessment practices illustrate the relevance and importance of a teacher's essential understanding of mathematics.

On behalf of the Board of Directors, I offer sincere thanks and appreciation to everyone who has helped to make this series possible. I extend special thanks to Rose Mary Zbiek for her leadership as series editor. I join the Essential Understanding project team in welcoming you to these books and in wishing you many years of continued enjoyment of learning and teaching mathematics.

Henry Kepner
President, 2008–2010

Preface

From prekindergarten through grade 12, the school mathematics curriculum includes important topics that are pivotal in students' development. Students who understand these ideas cross smoothly into new mathematical terrain and continue moving forward with assurance.

However, many of these topics have traditionally been challenging to teach as well as learn, and they often prove to be barriers rather than gateways to students' progress. Students who fail to get a solid grounding in them frequently lose momentum and struggle in subsequent work in mathematics and related disciplines.

The Essential Understanding Series identifies such topics at all levels. Teachers who engage students in these topics play critical roles in students' mathematical achievement. Each volume in the series invites teachers who aim to be not just proficient but outstanding in the classroom—teachers like you—to enrich their understanding of one or more of these topics to ensure students' continued development in mathematics.

How much do you need to know?

To teach these challenging topics effectively, you must draw on a mathematical understanding that is both broad and deep. The challenge is to know considerably more about the topic than you expect your students to know and learn.

Why does your knowledge need to be so extensive? Why must it go above and beyond what you need to teach and your students need to learn? The answer to this question has many parts.

To plan successful learning experiences, you need to understand different models and representations and, in some cases, emerging technologies as you evaluate curriculum materials and create lessons. As you choose and implement learning tasks, you need to know what to emphasize and why those ideas are mathematically important.

While engaging your students in lessons, you must anticipate their perplexities, help them avoid known pitfalls, and recognize and dispel misconceptions. You need to capitalize on unexpected classroom opportunities to make connections among mathematical ideas. If assessment shows that students have not understood the material adequately, you need to know how to address weaknesses that you have identified in their understanding. Your understanding must be sufficiently versatile to allow you to represent the mathematics in different ways to students who don't understand it the first time.

In addition, you need to know where the topic fits in the full span of the mathematics curriculum. You must understand where your students are coming from in their thinking and where they are heading mathematically in the months and years to come.

Accomplishing these tasks in mathematically sound ways is a tall order. A rich understanding of the mathematics supports the varied work of teaching as you guide your students and keep their learning on track.

How can the Essential Understanding Series help?

The Essential Understanding books offer you an opportunity to delve into the mathematics that you teach and reinforce your content knowledge. They do not include materials for you to use directly with your students, nor do they discuss classroom management, teaching styles, or assessment techniques. Instead, these books focus squarely on issues of mathematical content—the ideas and understanding that you must bring to your preparation, in-class instruction, one-on-one interactions with students, and assessment.

How do the authors approach the topics?

For each topic, the authors identify "big ideas" and "essential understandings." The big ideas are mathematical statements of overarching concepts that are central to a mathematical topic and link numerous smaller mathematical ideas into coherent wholes. The books call the smaller, more concrete ideas that are associated with each big idea *essential understandings*. They capture aspects of the corresponding big idea and provide evidence of its richness.

The big ideas have tremendous value in mathematics. You can gain an appreciation of the power and worth of these densely packed statements through persistent work with the interrelated essential understandings. Grasping these multiple smaller concepts and through them gaining access to the big ideas can greatly increase your intellectual assets and classroom possibilities.

In your work with mathematical ideas in your role as a teacher, you have probably observed that the essential understandings are often at the heart of the understanding that you need for presenting one of these challenging topics to students. Knowing these ideas very well is critical because they are the mathematical pieces that connect to form each big idea.

How are the books organized?

Every book in the Essential Understanding Series has the same structure:

- The introduction gives an overview, explaining the reasons for the selection of the particular topic and highlighting some of the differences between what teachers and students need to know about it.

Big ideas and essential understandings are identified by icons in the books.

marks a big idea, and

marks an essential understanding.

- Chapter 1 is the heart of the book, identifying and examining the big ideas and related essential understandings.

- Chapter 2 reconsiders the ideas discussed in chapter 1 in light of their connections with mathematical ideas within the grade band and with other mathematics that the students have encountered earlier or will encounter later in their study of mathematics.

- Chapter 3 wraps up the discussion by considering the challenges that students often face in grasping the necessary concepts related to the topic under discussion. It analyzes the development of their thinking and offers guidance for presenting ideas to them and assessing their understanding.

The discussion of big ideas and essential understandings in chapter 1 is interspersed with questions labeled "Reflect." It is important to pause in your reading to think about these on your own or discuss them with your colleagues. By engaging with the material in this way, you can make the experience of reading the book participatory, interactive, and dynamic.

Reflect questions can also serve as topics of conversation among local groups of teachers or teachers connected electronically in school districts or even between states. Thus, the Reflect items can extend the possibilities for using the books as tools for formal or informal experiences for in-service and preservice teachers, individually or in groups, in or beyond college or university classes.

marks a "Reflect" question that appears on a different page.

A new perspective

The Essential Understanding Series thus is intended to support you in gaining a deep and broad understanding of mathematics that can benefit your students in many ways. Considering connections between the mathematics under discussion and other mathematics that students encounter earlier and later in the curriculum gives the books unusual depth as well as insight into vertical articulation in school mathematics.

The series appears against the backdrop of *Principles and Standards for School Mathematics* (NCTM 2000), *Curriculum Focal Points for Prekindergarten through Grade 8 Mathematics: A Quest for Coherence* (NCTM 2006), *Focus in High School Mathematics: Reasoning and Sense Making* (NCTM 2009), and the Navigations Series (NCTM 2001–2009). The new books play an important role, supporting the work of these publications by offering content-based professional development.

The other publications, in turn, can flesh out and enrich the new books. After reading this book, for example, you might select hands-on, Standards-based activities from the Navigations books for your students to use to gain insights into the topics that the Essential Understanding books discuss. If you are teaching students in prekindergarten through grade 8, you might apply your deeper understanding as you present material related to the three focal points that Curriculum Focal Points identifies for instruction at your students' level. Or if you are teaching students in grades 9–12, you might use your understanding to enrich the ways in which you can engage students in mathematical reasoning and sense making as presented in *Focus in High School Mathematics*.

An enriched understanding can give you a fresh perspective and infuse new energy into your teaching. We hope that the understanding that you acquire from reading the book will support your efforts as you help your students grasp the ideas that will ensure their mathematical success.

The authors of the current volume express their sincere appreciation to those who reviewed the manuscript at an earlier stage, including Leslie M. Brickner, Joel Haack, Rhoda Inskeep, Gladis Kersaint, Sharon Lewandowski, and Diane Thiessen. The suggestions and insights that they shared were very helpful in shaping the authors' ideas. The authors also extend appreciation and thanks to Rose Zbiek and Ed Rathmell for their support, ideas, patience, and encouragement during the preparation of this book.

Introduction

This book focuses on ideas about multiplication and division. These are ideas that you need to understand thoroughly and be able to use flexibly to be highly effective in your teaching of mathematics in grades 3–5. The book discusses many mathematical ideas that are common in elementary school curricula, and it assumes that you have had a variety of mathematics experiences that have motivated you to delve into—and move beyond—the mathematics that you expect your students to learn.

The book is designed to engage you with these ideas, helping you to develop an understanding that will guide you in planning and implementing lessons and assessing your students' learning in ways that reflect the full complexity of multiplication and division. A deep, rich understanding of ideas about these operations will enable you to communicate their influence and scope to your students, showing them how these ideas permeate the mathematics that they have encountered—and will continue to encounter—throughout their school mathematics experiences.

The understanding of multiplication and division that you gain from this focused study thus supports the vision of *Principles and Standards for School Mathematics* (NCTM 2000): "Imagine a classroom, a school, or a school district where all students have access to high-quality, engaging mathematics instruction" (p. 3). This vision depends on classroom teachers who "are continually growing as professionals" (p. 3) and routinely engage their students in meaningful experiences that help them learn mathematics with understanding.

Why Multiplication and Division?

Like the topics of all the volumes in NCTM's Essential Understanding Series, multiplication and division compose a major area of school mathematics that is crucial for students to learn but challenging for teachers to teach. Students in grades 3–5 need to understand these operations well if they are to succeed in these grades and in their subsequent mathematics experiences. Learners often struggle with ideas about multiplication and division. What is the relationship between multiplication and repeated addition, for example? Many students understand multiplication only as repeated addition. The importance of understanding how arithmetic operations relate to one another and the challenge of computing fluently make it essential for teachers of grades 3–5 to understand multiplication and division extremely well themselves.

Your work as a teacher of mathematics in these grades calls for a solid understanding of the mathematics that you—and your school, your district, and your state curriculum—expect your students to learn about multiplication and division. Your work also requires you to know how this mathematics relates to other mathematical ideas that your students will encounter in the lesson at hand, the current school year, and beyond. Rich mathematical understanding guides teachers' decisions in much of their work, such as choosing tasks for a lesson, posing questions, selecting materials, ordering topics and ideas over time, assessing the quality of students' work, and devising ways to challenge and support their thinking.

Understanding Multiplication and Division

Teachers teach mathematics because they want others to understand it in ways that will contribute to success and satisfaction in school, work, and life. Helping your students develop a robust and lasting understanding of multiplication and division requires that you understand this mathematics deeply. But what does this mean?

It is easy to think that understanding an area of mathematics, such as multiplication and division, means knowing certain facts, being able to solve particular types of problems, and mastering relevant vocabulary. For example, for the upper elementary grades, you are expected to know such facts as "multiplication of whole numbers is a commutative operation." You are expected to be skillful in solving problems that involve multiplying large numbers. Your mathematical vocabulary is assumed to include such terms as *product*, *divisor*, *remainder*, *factor*, and *multiple*.

Obviously, facts, vocabulary, and techniques for solving certain types of problems are not all that you are expected to know about multiplication and division. For example, in your ongoing work with students, you have undoubtedly discovered that you need not only to know common algorithms for multiplication and division but also to be able to follow strategies that your students create.

It is also easy to focus on a very long list of mathematical ideas that all teachers of mathematics in grades 3–5 are expected to know and teach about multiplication and division. Curriculum developers often devise and publish such lists. However important the individual items might be, these lists cannot capture the essence of a rich understanding of the topic. Understanding multiplication and division deeply requires you not only to know important mathematical ideas but also to recognize how these ideas relate to one another. Your understanding continues to grow with experience and as a result of opportunities to embrace new ideas and find new connections among familiar ones.

Furthermore, your understanding of multiplication and division should transcend the content intended for your students. Some of the differences between what you need to know and what you expect them to learn are easy to point out. For instance, your understanding of the topic should include a grasp of the way in which multiplication of two-digit numbers connects with multiplication of binomial expressions—mathematics that students will encounter later but do not yet understand.

Other differences between the understanding that you need to have and the understanding that you expect your students to acquire are less obvious, but your experiences in the classroom have undoubtedly made you aware of them at some level. For example, how many times have you been grateful to have an understanding of multiplication and division that enables you to recognize the merit in a student's unanticipated mathematical question or claim? How many other times have you wondered whether you could be missing such an opportunity or failing to use it to full advantage because of a gap in your knowledge?

As you have almost certainly discovered, knowing and being able to do familiar mathematics are not enough when you're in the classroom. You also need to be able to identify and justify or refute novel claims. These claims and justifications might draw on ideas or techniques that are beyond the mathematical experiences of your students and current curricular expectations for them. For example, you may need to be able to refute the often-asserted, erroneous claim that three or more whole numbers can be added, subtracted, multiplied, or divided in any order. Or you may need to explain to a student why "carrying" is appropriate in some multiplication work.

Big Ideas and Essential Understandings

Thinking about the many particular ideas that are part of a rich understanding of multiplication and division can be an overwhelming task. Articulating all of those mathematical ideas and their connections would require many books. To choose which ideas to include in this book, the authors considered a critical question: What is essential for teachers of mathematics in grades 3–5 to know about multiplication and division to be effective in the classroom? To answer this question, the authors drew on a variety of resources, including personal experiences, the expertise of colleagues in mathematics and mathematics education, and the reactions of reviewers and professional development providers, as well as ideas from curricular materials and research on mathematics learning and teaching.

As a result, the mathematical content of this book focuses on essential ideas for teachers about multiplication and division. In particular, chapter 1 is organized around two big ideas related to this important area of mathematics. Each big idea is supported by smaller, more specific mathematical ideas, which the book calls *essential understandings*.

Benefits for Teaching, Learning, and Assessing

Understanding multiplication and division can help you implement the Teaching Principle enunciated in *Principles and Standards for School Mathematics*. This Principle sets a high standard for instruction: "Effective mathematics teaching requires understanding what students know and need to learn and then challenging and supporting them to learn it well" (NCTM 2000, p. 16). As in teaching about other critical topics in mathematics, teaching about multiplication and division requires knowledge that goes "beyond what most teachers experience in standard preservice mathematics courses" (p. 17).

Chapter 1 comes into play at this point, offering an overview of multiplication and division that is intended to be more focused and comprehensive than many discussions of the topic that you are likely to have encountered. This chapter enumerates, expands on, and gives examples of the big ideas and essential understandings related to the operations, with the goal of supplementing or reinforcing your understanding. Thus, chapter 1 aims to prepare you to implement the Teaching Principle fully as you provide the support and challenge that your students need for robust learning about multiplication and division.

Consolidating your understanding in this way also prepares you to implement the Learning Principle outlined in *Principles and Standards*: "Students must learn mathematics with understanding, actively building new knowledge from experience and prior knowledge" (NCTM 2000, p. 20). To support your efforts to help your students learn about multiplication and division in this way, chapter 2 builds on the understanding of these operations that chapter 1 communicates by pointing out specific ways in which the big ideas and essential understandings connect with mathematics that students typically encounter earlier or later in school. This chapter supports the Learning Principle by emphasizing longitudinal connections in students' learning about multiplication and division. For example, as their mathematical experiences expand, students gradually develop an understanding of the connections between multiplication and division of whole numbers and the corresponding operations with fractions and decimals and become fluent in using properties of operations and numbers in estimation and mental mathematics.

The understanding that chapters 1 and 2 convey can strengthen another critical area of teaching. Chapter 3 addresses this area, building on the first two chapters to show how an understanding of multiplication and division can help you select and develop appropriate tasks, techniques, and tools for assessing your students' understanding of the operations. An ownership of the big ideas and essential understandings related to multiplication and division, reinforced by an understanding of students' past and future experiences with the ideas, can help you ensure that assessment in your classroom supports the learning of significant mathematics.

Such assessment satisfies the first requirement of the Assessment Principle set out in *Principles and Standards*: "Assessment should support the learning of important mathematics and furnish useful information to both teachers and students" (NCTM 2000, p. 22). An understanding of multiplication and division can also help you satisfy the second requirement of the Assessment Principle, by enabling you to develop assessment tasks that give you specific information about what your students are thinking and what they understand. For example, when asked to multiply two two-digit numbers such as 24 and 35, a student might write something like

$$\begin{array}{r} 2 \\ 35 \\ \times\, 24 \\ \hline 140 \end{array}$$

while saying "4 times 5 is 20, so I put down the zero and carry the 2; 4 times 3 is 12 and 2 more is 14, so I write down 140; and then...." The student's response to the question, "Why do you add 12 and 2 when this is a multiplication problem?" can reveal much about how the student understands the mechanics of a common algorithm that he or she will soon extend to products of numbers with more than two digits.

Ready to Begin

This introduction has painted the background, preparing you for the big ideas and associated essential understandings related to multiplication and division that you will encounter and explore in chapter 1. Reading the chapters in the order in which they appear can be a very useful way to approach the book. Read chapter 1 in more than one sitting, allowing time for reflection. Absorb the ideas—both big ideas and essential understandings—related to multiplication and division. Appreciate the connections among these ideas. Carry your newfound or reinforced understanding to chapter 2, which guides you in seeing how the ideas related to these operations are connected to the mathematics that your students have encountered earlier or will encounter later in school. Then read about teaching, learning, and assessment issues in chapter 3.

Alternatively, you may want to take a look at chapter 3 before engaging with the mathematical ideas in chapters 1 and 2. Having the challenges of teaching, learning, and assessment issues clearly in mind, along with possible approaches to them, can give you a different perspective on the material in the earlier chapters.

No matter how you read the book, let it serve as a tool to expand your understanding, application, and enjoyment of multiplication and division.

with the number of terms, or addends (5), representing the number of shirts. A common description of such a situation is "5 groups of 4 objects," where the objects are the buttons on *one* shirt.

We might also write this sum as the multiplicative expression 5 × 4. Multiplication is sometimes described as repeated addition. Unfortunately, this description can create some rather awkward interpretations in other situations—for example, with fractions, as chapter 2 will discuss.

Under an interpretation of multiplication as a scalar operation, the total number of buttons is five times as many as the number of buttons on one shirt. We could say that the 5 is prescribing the multiplicative action that we should make with the number of buttons on the one shirt. We want 5 times as many buttons. Note that we have not calculated the numerical value of 5 × 4, since our focus is on the meanings of the factors 5 and 4.

We could also represent the situation by using the multiplicative expression 4 × 5, without using the commutative property, by noting that we have 4 groups of 5 objects. We have 5 first buttons on the 4 shirts, 5 second buttons on the 4 shirts, and so on, for the other two buttons. Interpreting the situation in this way, as "4 groups of 5 objects," might not be as natural a choice as interpreting it in the first way, as "5 groups of 4."

An explanation is always necessary to clarify the reasoning behind an expression or equation. Note that in the first interpretation of the problem, as 5 groups of 4 objects, with the accompanying expression 5 × 4, we have been very careful to use 5 to describe the number of shirts, not the shirts themselves. Similarly, we have used 4 to refer to the number of buttons on a shirt, not the buttons themselves. In the second interpretation of the problem, as 4 groups of 5 objects, with the accompanying expression 4 × 5, 4 represents the number of buttons, not the buttons themselves, and 5 describes, for example, the number of first buttons, not the first buttons themselves.

Making this kind of distinction is an important part of establishing understanding in mathematics, especially algebra. Other numbers besides 4 might be associated with the buttons. For example, 25 might be their cost in cents, and 2 might be their weight in ounces. Dougherty and colleagues (2010) offer an extended discussion of the important difference between objects and quantities.

A picture showing 5 shirts, each with 4 buttons, is not, of course, the only possible representation of the situation. Figure 1.3 shows a table with five rows, each of which represents a shirt, and four columns, each of which represents a button. In the table, each cell represents a certain button on a certain shirt. For example, the orange cell in the figure represents the third button on the second shirt. The rows and columns of this table represent the situation in example 2 as an *array*.

For an extended discussion of how different quantities can be associated with an object, see *Developing Essential Understanding of Number and Numeration for Teaching Mathematics in Prekindergarten–Grade 2* (Dougherty et al. 2010).

shirt\button	button 1	button 2	button 3	button 4
shirt 1				
shirt 2			shirt 2 button 3	
shirt 3				
shirt 4				
shirt 5				

Fig. 1.3. A table representing five shirts, each with four buttons

The use of an array as a representation is more familiar in the context of a situation like that in example 3:

Example 3: A classroom has a rectangular arrangement of desks with 5 rows of 4 desks. How many desks are in the classroom?

A natural pictorial representation would be a rectangular array with 5 horizontal rows, with each row repeating the same symbol four times, to represent the desks in the row. Figure 1.4 shows one example. Reflect 1.3 provides an opportunity to make sense of the representation and situation.

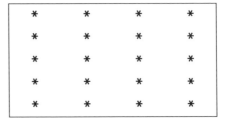

Fig. 1.4. Representing 5 rows of 4 desks as an array

Reflect 1.3

How does 4 × 5 make sense as an answer to the question in example 3?

How does 5 × 4 make sense as an answer to the question?

Representing situations with multiplication and division

Essential Understanding 1c. *A situation that can be represented by multiplication has an element that represents the scalar and an element that represents the quantity to which the scalar applies.*

Problem solvers often choose the operation of division for use in solving a problem such as example 4:

Example 7 presents a problem that involves finding the area of a rectangular region:

Example 7: A rectangular carpet measures 5 feet by 4 feet. How many square feet of floor will the carpet cover?

A typical strategy for solving this problem would be to multiply the length and the width to get an answer of 20 square feet. Why is this situation multiplicative?

We might represent the carpet situation by drawing a rectangle in which one side has a length of 5 units and another side has a length of 4 units, as shown in figure 1.8. Then we might partition the rectangle into unit squares, and so the answer would be the number of unit squares, or, in example 7, 20 square feet.

Fig. 1.8. A rectangle representing carpet that is 5 feet by 4 feet

Consider only the first row of the representation. It is a rectangle with one row of unit squares. This first row with 4 unit squares has area 4. The entire representation includes 5 rows, which means there are 5 times as many rows, or, equivalently, 5 times as many unit squares as in the first row. This is a multiplicative situation in which 5 is a scalar operator indicating that we want 5 times as much area as the area of the first row. The number of unit squares, or area, is expressed by 5 × 4.

Moving beyond operations on whole numbers

Essential Understanding 1d. *A scalar definition of multiplication is useful in representing and solving problems beyond whole number multiplication and division.*

A scalar interpretation of multiplication can give students broad and flexible understanding of multiplication and division and help them reason about and explain their use of these operations in problem solving. Example 8 offers a starting point for examining this idea:

> **Example 8:** Write an algebraic expression to show that Tom has twice as many pounds of candy as Susan.

Representing the situation with a diagram, followed by carefully and precisely defining the variables or components of that diagram, is a good opening strategy. When students attempt to use algebraic strategies to solve word problems, a failure to describe variables carefully at the start often dooms their efforts. For example, T can represent the number of pounds of candy that Tom has, and S can represent the number of pounds of candy that Susan has, but students sometimes fail to note that both T and S represent numbers of pounds—not the candy itself. A solution needs to describe the relationship between the amount of candy that Tom has and the amount of candy that Susan has.

Students often attempt to translate the words of a problem directly into a symbolic expression. In the case of example 8, this might lead them to write $2T = S$. Interpreting the multiplication expression $2T$ by taking 2 as a scalar factor indicates that if the amount of candy that Tom has is doubled, then this doubled amount would be the same amount of candy that Susan has. But this is obviously not the situation described in the example.

The diagram in figure 1.9 shows the relationship between the amounts of candy that Susan and Tom have. The first figure represents the amount of candy Susan has (S), and the second figure represents the amount of candy Tom has (T). Now it is clear that if the amount that Susan has is doubled, as represented by $2S$, it would then be the same as the amount that Tom has. Hence, the relationship between the amounts of candy that Tom and Susan have is $T = 2S$.

Fig. 1.9. A diagram showing that Tom has twice as many pounds of candy as Susan

Multiplication as a scalar operation provides a way of thinking about the situation and a justification for the choice of multiplication. Again, the importance of providing a justification for the choice of an operation is an important part of understanding and doing mathematics. Furthermore, the advantage of establishing a robust understanding of the problem, such as that provided by a pictorial representation, rather than beginning with the manipulation of algebraic symbols, is apparent.

Example 9 offers another glimpse of the value of the scalar interpretation in reasoning about and explaining the choice of operation in a solution strategy:

Example 9: Jose drives at an average speed of 50 miles per hour for 6 hours. How many miles does Jose drive?

This situation is generally recognized as a multiplicative one calling for the use of the distance = rate × time formula, or $d = rt$. Following the order of the letters labeling the variables in the formula, most students would write 50 × 6 as the multiplication expression that will give the solution. But the format of this answer suggests that we are looking at 50 groups of 6 objects. What would each object be? An hour? Also, why do we use the operation of multiplication? Examining this problem from a different perspective offers an alternate way to make sense of the situation.

Jose drives at an average a rate of 50 miles per hour, which we can think of as meaning that Jose travels 50 miles in one hour and in each hour thereafter. What happens when Jose drives 6 times as much as he does in one hour? As suggested by the previous example, a pictorial representation can establish an understanding of the problem. In figure 1.10, each line segment represents 50 miles, which is the distance Jose travels in one hour, and the six segments represent Jose's travel in six hours. Multiplication as a scalar operation applies when the 6 is interpreted as a scalar describing how the quantity 50 is changed, yielding a multiplicative expression of 6 × 50. The answer and the initial quantity—the quantity to which the scalar applies—have the same units—namely, miles.

Fig. 1.10. Six segments, with each representing 50 miles traveled in one hour

We can use multiplication to develop and generalize a relationship among distance, rate, and time. We begin by defining the variables carefully and precisely:

D represents the total number of miles traveled.
T represents the number of hours used in traveling D miles.
R represents the number of miles traveled in each hour.

One way to make sense of this general situation is to reason as before, using a scalar view of multiplication. If we consider T as a scalar acting on R, the relationship makes sense. R is the number of miles traveled each hour, so $T \times R$ yields the number of miles traveled in T hours. This interpretation of the multiplicative situation leads to the algebraic formula $D = T \times R$. In any rate problem involving multiplication, it is then critical to consider what is being used as the scalar.

Every choice of an operation requires a justification: why is this operation appropriate? An understanding of the meaning of the operation is demonstrated in this way. Consider the following two examples:

Example 10: If an object weighs 156 pounds, and one pound has approximately the same weight as 0.454 kilograms, then what is an approximate weight of the object in kilograms?

This problem previously appeared in Reflect 1.2. Can you recall your ideas and reasoning as you considered the problem there? Compare your earlier thinking with the ideas that a scalar interpretation has given you now.

Example 11 presents a problem for which the operation of choice would probably be either multiplication or addition:

Example 11: Suppose that you are writing the first few terms in the sequence 1, 3, 6, 10, ... of triangular numbers. What is the next term that you will write?

The choice of an operation should be based on an understanding of the situation—in this case, the structure of the triangular numbers—and the meaning of the chosen operation. Looking at some noncontextual pattern or using some memorized technique, such as dimensional analysis, should not be a basis for the choice. Knowing which operation to use and why it is appropriate is an indicator of understanding.

A common aim in the previous examples has been to see how describing multiplication as a scalar operation makes sense for a variety of reasons. In a multiplicative expression, the first factor describes the action that is to happen to the quantity described by the second factor. The first number in the product will change, or *scale*, the quantity described by the second number in the expression. The change is multiplicative. This description is powerful and helps students to develop a robust understanding of multiplication. As a result of applying this interpretation, problem solvers can—

- describe many varied situations in a multiplicative manner;
- use multiple representations to describe multiplicative situations;
- provide justifications for using multiplication as an operation;
- make connections among different representations;
- trace the "answer," or product, of a multiplicative expression to the interpretation of a representation; and
- give meanings to the factors in a multiplicative expression.

Relating division and multiplication

Essential Understanding 1e. *Division is defined by its inverse relationship with multiplication.*

So far, our discussion has centered on multiplication and on a way to describe this operation that shows that it is an appropriate choice in many different situations. The discussions of example 4, in which Grandmother has 21 marbles and gives 7 to each of her grandchildren, and example 5, in which Grandfather has 24 old coins to share equally among his 6 grandchildren, make brief references to the use of division in finding the number of equal-sized groups and the number of items in equal-sized groups. These examples present situations in which the operation of division can be used. But that observation in itself does not give a meaning for the operation of division that can be used to justify the choice of division.

Consider the relationship between addition and subtraction. When b is subtracted from a, the result is represented by $a - b$ and is defined to be the number y where $a = b + y$. Addition and subtraction have an inverse relationship. In other words, addition is a fundamental operation and subtraction is defined in terms of addition. Likewise, multiplication is a fundamental operation, with division defined in terms of multiplication.

We give meaning to division in a manner that is analogous to the way in which we give meaning to subtraction. With $b \neq 0$, the result of dividing a by b is represented by $a \div b$ and is defined to be the number y where $a = b \times y$, or, equivalently, $y = a \times (1/b)$ where $1/b$ is the notation for the multiplicative inverse of b. Note the similar situation with addition and subtraction, in which we know that $a - b = a + (-b)$ where $-b$ is the notation for the additive inverse of a. The first format gives the connection of division with multiplication, so multiplication and division have an inverse relationship. The equation $a = b \times y$ describes the same relationships among the numbers as the equation $a \div b = y$. The expression $a \times (1/b)$ is often written as $a(1/b)$, or a/b, a connection with fractions that is consistent with the way in which fractions are used.

The inverse relationship between multiplication and division provides the mathematical basis for the fact families, such as the following:

$$7 \times 3 = 21 \qquad 3 \times 7 = 21 \qquad 21 \div 3 = 7 \qquad 21 \div 7 = 3.$$

Which of the four facts above we use depends on the situation and our reasoning for the choice of the operation. Chapter 3 discusses this thinking in more detail, and Caldwell and colleagues (2011) discuss similar thinking in relation to addition and subtraction.

In example 4, where Grandmother has 21 marbles and gives 7 to each grandchild, problem solvers must determine how many

For a discussion of the usefulness of addition fact families, see *Developing Essential Understanding of Addition and Subtraction for Teaching Mathematics in Prekindergarten–Grade 2* (Caldwell et al. 2011).

grandchildren Grandmother has. As we noted earlier, problem solvers frequently choose to use division in solving this kind of problem, but often without being able to justify the choice. Indeed, when we discussed this problem, we had not yet discussed the mathematical reasons that justify the choice of division, so instead we employed a strategy using multiplication to determine the number of grandchildren, A, by developing the algebraic equation $A \times 7 = 21$. Now, however, by using our recognition of the inverse relationship between multiplication and division, we can write the equation $A = 21 \times (1/7)$. This equation indicates that we could have obtained the answer by using division, dividing 21 by 7. Thus, the choice of division is fully appropriate, since we have now given a justification for it. Reflect 1.6 probes the use of similar reasoning to justify the choice of division in the case of example 5.

Reflect 1.6

Explain why division can be an appropriate choice of operation to use in solving the problem in example 5:

Grandfather has 24 old coins. He is going to give each of his 6 grandchildren the same number of coins. How many will each grandchild receive?

Using numbers in division situations

Essential Understanding 1f. *Using proper terminology and understanding the division algorithm provide the basis for understanding how numbers such as the* quotient *and the* remainder *are used in a division situation.*

Often the terminology that is used with division is vague, inappropriate, or misleading. It can even interfere with the development of a rich understanding of division. Proper use of terminology and symbols can increase understanding of the meanings of arithmetic expressions and help to avoid mistakes. Consider the following examples:

- "Suppose that a group of apples is divided in half." The meaning of this statement is ambiguous, as illustrated in parts (a) and (b) of figure 1.11. Does it mean that the group of apples is partitioned into two equal-sized groups of whole apples, as shown in (a), or does it mean that each apple is partitioned into two same-sized pieces, as shown in (b)? These are two very different situations, as the figure suggests.

- "How many times does 6 go into 42?" The use of the phrase "goes into" is frequently used in conjunction with division,

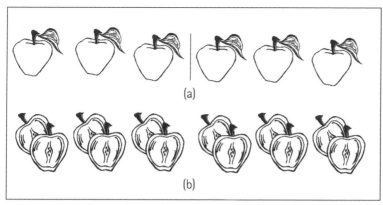

Fig. 1.11. Six apples (a) partitioned into two equal-sized groups of whole apples and (b) with each apple partitioned into two same-size pieces

particularly long division. This phrase really has no mathematical meaning, so it should be avoided.

- "Suppose that 6 is divided into 12." This phrase lends itself to various interpretations. Does it mean to divide 6 by 12 or 12 by 6? Does it mean to do the division, or does it refer to the answer? Phrases that can be interpreted in different ways should be avoided; otherwise the listener—usually a student—can interpret them in ways that are not intended.

The use of correct notation to represent some aspect of a problem is often a very important first step. For example, students frequently encounter statements such as, "The natural number a is divisible by 3." They may then proceed to represent this statement by the expression $a/3$, which is an incorrect representation and often makes finding a solution strategy very difficult, if one can be found at all. It would appear that students are confusing the *result* of dividing a by 3 with the *operation* of dividing a by 3. These are two separate ideas. In general, when b is a nonzero divisor, it is appropriate to say "divide a by b," with the result written as a/b. Note that a/b is an answer to a division problem and does not mean to divide a by b or that a is divisible by b. The operation of division has already been carried out when we see an expression such as a/b. Explaining answers to the questions in Reflect 1.7 can help to connect notation and terminology for division and multiplication.

Reflect 1.7

We can say "divide a by b."
Is it appropriate to say "divide a and b"?
Does it make sense to say "multiply $a \times b$"?
Does it make sense to say "divide a/b"?

It is critical that mathematical ideas be communicated carefully and stated precisely to promote an understanding of concepts, relationships, and representations. Mathematical terms are well defined and should be used instead of words and phrases that do not have mathematical meanings.

The use of the term *divisible* involves a relationship between two whole numbers. The whole number A is divisible by the natural number B when there is a whole number Y such that $A = B \times Y$. A is divisible by B when the division of A by B results in a remainder of 0. For example, 15 is divisible by 5, but 12 is not divisible by 5, and 0 is divisible by 7 and every other natural number. Other ways of saying that A is divisible by B include saying that B is a factor of A or that A is a multiple of B.

It can be misleading to say, "B goes into A evenly." The use of the word *evenly* may evoke the idea that even numbers are involved, although they are not. There is really no reason to use *evenly* when the mathematically precise terms *multiple* and *factor* can be used.

After dividing, people sometimes say, "There is no remainder." That statement is not true. There is always a remainder after division, and those who make this statement probably mean that the remainder is 0. Having a remainder of 0 is not the same as having "no remainder."

In fact, *remainder* and *quotient* are two other important ideas associated with the concept of division. When a whole number is divided by a natural number, it may or may not be that the *dividend* is a multiple of the *divisor*. In general, when a whole number a is divided by a natural number b, the result is a number q, called the *quotient*, and a whole number r, called the *remainder*, such that $a = bq + r$ where $0 \leq r < b$. It should be noted that q and r are the only numbers that satisfy these conditions; that is, they are said to be *unique*. This statement is called the *division algorithm*. For example, the division relationship between 13 and 5 is $13 = (2 \times 5) + 3$. Note that 2 is the quotient, and 3 is the remainder, which is less than the divisor, 5. Bates and Rousseau (1986) provide a number of ways to represent the division algorithm. Reflect 1.8 explores the use of a number line to justify the uniqueness of the quotient and remainder.

In many situations, problem solvers need to understand how the divisor, the quotient, and the remainder can be used in the context. Even when they have calculated the quotient and remainder, they need to interpret each of them in the context of the situation and understand how to use each to find an answer. Consider the questions in example 12:

See Reflect 1.8
on p. 27.

Example 12: Suppose that Tina has 43 oranges that she is packing in into identical empty boxes. Tina experiments and discovers that 8 oranges completely fill a box.

second interpretation yielded the multiplicative expression 4×5.

Because the number of buttons is the same in either interpretation, it makes sense to write the equality $5 \times 4 = 4 \times 5$, which is an instance of the commutative property of multiplication. Although 5×4 and 4×5 have the same numerical value, our interpretations of the two are not the same. With 5×4, 5 is the scalar acting on 4, the number of buttons on one shirt. With 4×5, 4 is the scalar acting on the number of "first buttons" (or nth buttons) on the 5 shirts.

Students quickly realize that the commutative property can make calculations easier, even if they do not know the property by name or use it deliberately. For example, to compute 8×2, (8 groups of 2, or $2 + 2 + 2 + 2 + 2 + 2 + 2 + 2$), some students sometimes find it easier and more efficient to find 2×8 (2 groups of 8 or $8 + 8$).

An understanding of the commutative property can make nearly half of the basic multiplication facts easier to learn. In the table in figure 1.12, note that 4×2 is directly across the main diagonal from 2×4. In fact, because the products $a \times b$ and $b \times a$ are the same for all pairs of numbers a and b, these products appear in squares that are symmetric with respect to the main diagonal. Because of the commutative property, students learning multiplication facts need to learn only the facts represented by the orange and white cells in the table in figure 1.12.

×	0	1	2	3	4	5	6	7	8	9
0	0	0	0	0	0	0	0	0	0	0
1	0	1	2	3	4	5	6	7	8	9
2	0	2	4	6	8	10	12	14	16	18
3	0	3	6	9	12	15	18	21	24	27
4	0	4	8	12	16	20	24	28	32	36
5	0	5	10	15	20	25	30	35	40	45
6	0	6	12	18	24	30	36	42	48	54
7	0	7	14	21	28	35	42	49	56	63
8	0	8	16	24	32	40	48	56	64	72
9	0	9	18	27	36	45	54	63	72	81

Fig. 1.12. The symmetry in the multiplication table reflects the commutative property of multiplication for whole numbers.

For division, however, as for subtraction (see Caldwell et al. [2011]), order *does* matter. The following quotients are different:

$$12 \div 3 = 4 \qquad \text{and} \qquad 3 \div 12 = {}^1/_4$$

One counterexample is sufficient to show that a property does not hold for the system of whole numbers, so we can say that while multiplication and addition are commutative, division and subtraction are not. Reflect 1.10 probes the relationship between the different quotients when the dividend and the divisor are reversed.

Reflect 1.10

Explain how the following two quotients are related:

$$12 \div 3 = 4 \qquad \text{and} \qquad 3 \div 12 = 1/4$$

Compare this relationship to that between the differences when the minuend and subtrahend are reversed in subtraction:

$$7 - 4 = 3 \qquad \text{and} \qquad 4 - 7 = -3$$

Associative property: $(a \times b) \times c = a \times (b \times c)$

Multiplicative expressions such as $7 \times 6 \times 5$ are commonplace. However, we can multiply only two numbers at a time—in this case, either 7 and 6 or 6 and 5. So, which one of the two possibilities should we multiply first? We see that we can use parentheses to write $7 \times 6 \times 5$ in two different ways: $(7 \times 6) \times 5$ and $7 \times (6 \times 5)$. Both have the same value, but 7×30 allows an easier mental calculation than 42×5. The associative property for multiplication permits us to write the multiplication of three or more whole numbers without using any parentheses and still get the same result, regardless of how we group the factors. For example, what are some different ways to compute $2 \times 3 \times 4 \times 5$ without using the commutative property?

To help understand why the associative property for multiplication holds for whole numbers, we can think about the volume of a rectangular prism with edges of lengths 5, 6, and 7. First, let's consider as the base the face with edges of lengths 6 and 5 (see fig. 1.13a). We can imagine the volume of the prism as composed of layers, each with a "thickness" of 1, stacked on this base. The bottom layer is a $(6 \times 5) \times 1$ rectangular prism with a volume of 6×5 cubic units. Our prism has 7 layers, resulting in a volume of $7 \times (6 \times 5)$ cubic units. We can turn the prism so that the face with sides of length 6 and 7 is the base (see fig. 1.13b). We now consider the volume as being composed of $1 \times 1 \times 5$ vertical rods, each with a volume of 5 cubic units. We have 7×6 rods, resulting in a volume of $(7 \times 6) \times 5$ cubic units.

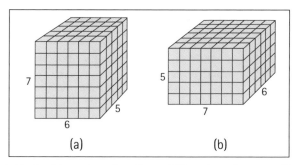

Fig. 1.13. Looking at a prism from different angles to verify the associative property of multiplication of whole numbers

We see that the number of cubic units is the same no matter how we compute the volume, so it makes sense that these two expressions are equivalent—an instance of the associative property for multiplication of whole numbers. We could develop the general statement of the associative property of multiplication for whole numbers in the same way by looking at a rectangular prism with sides of lengths a, b, and c.

What happens, however, if we consider the idea of associativity with respect to the operation of division? Let's look at an example:

$$(60 \div 6) \div 2 = 10 \div 2 = 5$$
$$60 \div (6 \div 2) = 60 \div 3 = 20$$

These values are different; hence, just as the operation of subtraction is not associative, so also the operation of division is not associative. How to evaluate a numerical expression such as $60 \div 6 \div 2$ is a matter that we will consider later in connection with order of operations.

The power of the commutative and associative properties of multiplication is in the flexibility that they afford in calculations. For example, students may compute the basic fact 4×7 sometimes by thinking of it as $(2 \times 2) \times 7$ and at other times by thinking of it as $2 \times (2 \times 7)$, options available to them by an application of the associative property. Flexibility in the use of the properties frequently depends on decomposing at least one of the numbers.

Consider the expression $8 \times 17 \times 25$. We can introduce parentheses in either of two places, writing $(8 \times 17) \times 25$ or $8 \times (17 \times 25)$. However, neither grouping provides us with an easy mental calculation. Nevertheless, by using what most upper elementary students know, together with the flexibility that the associative and commutative properties make possible, we can make the computation much simpler. Most students know that four quarters equal a dollar, so they know that 4×25 is 100. That knowledge makes 8×25 a value that they can easily obtain mentally, since

$$8 \times 25 = (2 \times 4) \times 25 = 2 \times (4 \times 25) = 2 \times 100 = 200,$$

by an application of the associative property. This process suggests a strategy for mentally calculating $8 \times 17 \times 25$:

$(8 \times 17) \times 25$	$= (17 \times 8) \times 25$	By the commutative property
	$= 17 \times (8 \times 25)$	By the associative property
	$= 17 \times ((2 \times 4) \times 25)$	Using what we know: $8 = 2 \times 4$
	$= 17 \times (2 \times (4 \times 25))$	By the associative property
	$= 17 \times (2 \times 100)$	Using what we know: $4 \times 25 = 100$
	$= (17 \times 2) \times 100$	By the associative property
	$= 34 \times 100$	Using what we know: $17 \times 2 = 34$
	$= 3400$	Using what we know: $34 \times 100 = 3400$

In this instance, we use both the associative and the commutative properties, together with some factual knowledge. Understanding what properties to use and the flexibility that they provide in computational situations are important features of robust understanding. How would you use the commutative and associative properties to simplify the computation of $4 \times 8 \times 15$?

These properties can be extended for the multiplication of more than three numbers. The generalized associative and commutative properties for multiplication mean that factors can be regrouped and rearranged in any order. These properties can be applied to simplify many calculations.

Using the generalized associative property with the expression $5 \times 8 \times 6 \times 4$, we can insert parentheses in different ways and still get the same value, as illustrated below:

$$((5 \times 8) \times 6) \times 4 = (40 \times 6) \times 4 = 240 \times 4 = 960$$
$$5 \times ((8 \times 6) \times 4) = 5 \times (48 \times 4) = 5 \times 192 = 960$$
$$(5 \times 8) \times (6 \times 4) = 40 \times 24 = (10 \times 4) \times 24 = 10 \times (4 \times 24)$$
$$= 10 \times 96 = 960$$

These examples illustrate—but do not prove—that parentheses are not needed for a product with any number of factors. For additional discussion of generalizing and reasoning as we have done with the associative property, see Lannin and colleagues (forthcoming).

For a discussion of mathematical reasoning and generalizing, see *Developing Essential Understanding of Mathematical Reasoning for Teaching Mathematics in Prekindergarten–Grade 8* (Lannin et al. forthcoming).

Chapter 2

Connections: Looking Back and Ahead in Learning

Multiplication and division are powerful ideas. Their use in computations and problem solving in grades 3–5 is only part of their contribution to the curriculum and to students' knowledge. This chapter examines connections between the essential understandings described in chapter 1 and other mathematical topics, while considering the foundations of these ideas in the primary grades and their application in subsequent grades.

An understanding of the operation of multiplication builds on ideas about place value, addition, and skip counting. It aids in developing an understanding of fractions and percents and lays a foundation for future work with ratios and similarity. An understanding of algorithms and the use of properties of whole number multiplication and division to justify them provides a basis for developing fluency in rational number computations and algebraic manipulations.

Multiplication as a Fundamental Operation

Each of the mathematical topics in this chapter connects with the big ideas and essential understandings described in chapter 1. The scalar definition of multiplication (Essential Understanding 1a) and the properties of multiplication are evident in other mathematical topics that students encounter throughout their school years. From place value, addition, and skip counting to percent, ratio, and similarity, thinking about the fundamental operation of multiplication as a scalar process connects concepts, procedures, and problems.

Essential ←
Understanding 1a

In the multiplicative expression A × B, A can be defined as a scaling factor.

Place value and addition

The base-ten numeration system is fundamental to understanding numbers and any written or mental algorithm with whole numbers.

These are key topics in the primary grades, where students learn that 23 is 2 tens and 3 ones, which they later write in expanded notation as $2 \times 10 + 3 \times 1$. In this expression, 2 is the scaling factor that tells how many tens, and 3 is the scaling factor that tells how many ones. When students learn to add 23 and 35, they add each place value separately. The connections of multiplication to place value are integrated into the addition algorithm through the use of the distributive property, together with the commutative and associative properties for addition:

$23 + 35 = (2 \times 10 + 3) + (3 \times 10 + 5)$	Place value
$= (2 \times 10 + 3 \times 10) + (3 + 5)$	Addition properties
$= 5 \times 10 + 8$	Distributive property
$= 58$	Place value

Understanding the interplay of place value and the operations of addition and multiplication is a major part of developing a robust understanding of whole number operations and is critical to decomposing factors and combining partial products.

Division and subtraction

Division is defined in terms of multiplication in much the same way that subtraction is defined in terms of addition. In fact, the symbolic descriptions of subtraction and division are quite similar. We recall that $a - b = a + (-b)$, where $-b$ is the notation for the additive inverse (or opposite) of b, and $a \div b = a \times (1/b)$, where $1/b$ is the notation for the multiplicative inverse (or reciprocal) of b, with $b \neq 0$. Of course, b cannot be 0, because such a case would involve division by zero, which is undefined.

Alternate and equivalent descriptions for subtraction and division are, respectively, that $a - b$ is defined to be the number y when $a = b + y$ and that $a \div b$ is defined to be the number y when $a = b \times y$, with $b \neq 0$. The similarity in symbols reflects the similarities in defining subtraction and division in terms of addition and subtraction, respectively.

Skip counting

Skip counting is a worthwhile activity that many primary teachers use with their students, and it is often viewed as an example of using the operation of addition. However, it is also multiplicative in nature. Furthermore, many other important mathematical ideas are embedded in skip counting.

Consider skip counting by 6: 6, 12, 18, 24, and so forth. These numbers form a sequence, and each natural number corresponds to a term in the sequence. The first term is 1×6, the second term

Challenges: Learning, Teaching, and Assessing

Teachers need to develop understanding of the big ideas and essential understandings related to multiplication and division. Otherwise, they often use an instructional approach that does not promote a flexible, growing understanding in their students. Consider the following discussion between two teachers.

Ms. Avery and Ms. Wojak are discussing their beliefs about what it means for the students in each of their respective classrooms to have an understanding of the concepts and number facts related to multiplication and division. Ms. Avery believes that if students memorize their basic facts, they will be able to solve problems. She believes that a teacher's role is to help students get the correct answer and that getting the correct answer shows understanding. She explains her approach to Ms. Wojak in more detail:

> My students have been studying fact families so that they can
> see the relationship among the three numbers in multiplication
> and division facts. If I show them a triangular card with the
> numbers 3, 4, and 12 on it, with one number in each corner, they
> know that 3×4 is 12, that 4×3 is 12, that $12 \div 4$ is 3, and
> that $12 \div 3$ is 4. We play a game using fact-family cards in an
> "Around the World" format, and we talk about how the numbers
> on the card are related, just like the people in a family. The stu-
> dents come to understand multiplication and division through
> this game.

Ms. Wojak thinks that an approach that emphasizes memorization does not necessarily build in-depth understanding. She elaborates on her own very different approach:

> That's interesting. For my students to understand multiplication
> and division, I select a number triple such as 5, 6, and 30, and
> I create a word problem with those numbers, using a context
> familiar to my students so that they can make sense of the

→ **Essential Understanding 1c**
A situation that can be represented by multiplications has an element that represents the scalar and an element that represents the quantity to which the scalar applies.

numbers. For example, I might begin by saying, "If there are 5 children and each gets 6 balloons, how many balloons are there?" I want to determine if they think of five 6s and count by sixes or if they realize they can use the commutative property and think six 5s and then count by fives to get an answer. Or some may use the distributive property and known facts to reason, saying something like, "If I know three 5s are 15, then six 5s are 15 and 15, which is 30." Later in the lesson, I might ask, "If there are 30 cookies and 6 children, how many cookies can each child get if they each get the same number?" When students realize that division and multiplication have an inverse relationship and know that 5 × 6 is 30, they can find that 30 cookies shared fairly among 6 children is 5 cookies each. Then I am confident that they understand multiplication and division.

→ **Essential Understanding 2a**
The commutative and associative properties of multiplication and the distributive property of multiplication over addition ensure flexibility in computations with whole numbers and provide justifications for sequences of computations with them.

Note that this approach has the potential to help Ms. Wojak's students develop robust understanding—while allowing Ms. Wojak to assess their understanding in relation to several essential understandings. After giving her students the problem involving the 5 children with 6 balloons apiece, Ms. Wojak can observe them as they think about the elements in this multiplicative situation and identify one as the scalar and the other as the quantity on which the scalar acts (Essential Understanding 1c). She can note whether or not the students make use of the distributive property to compute flexibly (Essential Understanding 2a). She can then reverse the situation involving the number triple 5, 6, 30, presenting the students with the new problem of sharing 30 cookies fairly among 6 children, to see if they understand the inverse relationship between division and multiplication (Essential Understanding 1e).

→ **Essential Understanding 1e**
Division is defined by its inverse relationship with multiplication.

Ms. Wojak continues thinking about how she plans instruction for her students. She knows that whenever she is introducing a concept, she usually starts with a word problem to determine what mathematics the students know. She listens for different reasoning strategies. For the cookie problem, she thinks that one student might say, "I started with 30 cookies and gave each person 1. Then 6 were gone from the 30, and I had 24." The student might continue double counting: "2 groups of 6, 18 left; 3 groups of 6, 12 left; 4 groups of 6, 6 left; 5 groups of 6, 0 left. So each person gets 5 cookies." If at that point no one knew how to represent the situation symbolically, Ms. Wojak thinks she would write the equation 5 × 6 = 30 on the board, reading the symbol "×" as "groups of" if those were the words that the student used. This equation would mean that 5 acts as the scalar, or operates on the quantity 6 (Essential Understanding 1a). Ms. Wojak explains that she would encourage a variety of different solution strategies during the lesson (Essential Understanding 1b).

→ **Essential Understanding 1a**
In the multiplicative expression A × B, A can be defined as a scaling factor.

Ms. Wojak believes that to develop her students' understanding of ideas and facts related to multiplication and division, she must recognize their reasoning and help them to represent it symbolically, exactly following the thinking processes that they have used. She makes opportunities for students to share ideas during strategy sessions that elicit several different reasoning processes that represent most of the ways in which students in the class are thinking.

Ms. Wojak meets individually with students who are not yet using a strategy to help them learn to model a situation directly with pictures or manipulative materials of their choice. Moreover, Ms. Wojak knows that repeated subtraction is just a beginning division strategy for most students and that to make sense of larger numbers, they will also need to make sense of more sophisticated strategies, such as those that use the right distributive property (Essential Understanding 2*b*). Her goal is for students to develop a rich and growing understanding of the concepts and strategies related to multiplication and division and to develop an efficient algorithm based on those understandings.

Both Ms. Avery and Ms. Wojak have an idea of what *understanding* means. Each teacher believes strongly that her students must understand concepts and facts related to multiplication and division. Yet, students in the two classes will develop different mathematical understandings.

Ms. Avery's aim is for her students to develop symbolic proficiency, while Ms. Wojak is striving to help her students develop a more robust understanding of multiplication and division. To accomplish her goal, Ms. Wojak must have a deep understanding of the essential understandings so that she can respond to the different reasoning that students use to represent and solve problems and help them develop more efficient solution strategies.

Teaching for Robust Understanding

The scenario with Ms. Avery and Ms. Wojak brings up the issue of what each teacher believes is the meaning of *understanding*. For some teachers, an indicator of understanding might be proficiency with symbolic procedures. Such teachers might ask their students questions like, "What do you do next?" and, "Where do you write that number?"

As Ms. Avery and Ms. Wojak continue their discussion about understanding of multiplication and division, they begin to identify more specific indicators of understanding. When Ms. Avery asks a student to multiply two two-digit numbers, such as 24 and 35, for example, she expects the student to say (and do) something like, "OK, 4 times 5 is 20, so I put down the zero and carry the 2; 4 times 3 is 12 and 2 more is 14, so I write down 140, and then...." This

Essential
Understanding 1*b*

Each multiplicative expression developed in the context of a problem situation has an accompanying explanation, and different representations and ways of reasoning about a situation can lead to different expressions or equations.

Essential
Understanding 2*b*
The right distributive property of division over addition allows computing flexibly and justifying computations with whole numbers, but there is no left distributive property of division over addition and no commutative or associative property of division of whole numbers.

"explanation" does not provide evidence that a student has a rich understanding of the mathematical concepts involved in two-digit multiplication. It merely describes a symbolic procedure that might reflect a limited understanding.

Teachers like Ms. Wojak, by contrast, focus on the thinking that students are using, and they ask questions to help the students enhance their understanding as they develop other ways to represent and solve problems. These teachers' questions are designed to push behind procedures and probe students' understanding of their steps:

- Why do you put down a zero here?

- What does the 2 that you carried here represent?

- Why do you add 10 and 2, when this is a multiplication problem?

These types of questions address the "whys" of the process. They draw on the teacher's understanding of Big Idea 2 rather than his or her familiarity and efficiency with procedures. Furthermore, instead of eliciting from students a list of memorized procedural steps, these questions compel students to reason about, and build their understanding of, concepts. Teachers who focus on this kind of understanding of multi-digit multiplication help their students make connections among a variety of mathematical concepts. Let's examine some of the ways in which a teacher with deep understanding might help students learn to calculate a product like 24 × 35 with a flexible, robust, and growing understanding.

Big Idea 2

The properties of multiplication and addition provide the mathematical foundation for understanding computational procedures for multiplication and division, including mental computation and estimation strategies, invented algorithms, and standard algorithms.

Beginning with a word problem

A good way to begin is to create a word problem to assess strategies that students might intuitively use. For example, Mr. Robinson creates the following word problem and shares it with his students to gain insight into their understanding of the multiplication of two two-digit numbers:

> Each of 24 students in the fourth grade brought 75 cents for a school party. How much money did the fourth grade then have for the party?

Mr. Robinson observes that one student, Olivia, writes the following in her first attempt to solve the problem:

$$2 \times 75 = 150$$
$$4 \times 75 = 150 + 150$$
$$8 \times 75 = 300 + 300$$
$$16 \times 75 = 600 + 600 = 1200$$

Mr. Robinson sees that Olivia knows that she needs only 8 more groups of 75. Because 16 × 75 = 1200 and 8 × 75 = 600,

$24 \times 75 = 1200 + 600 = 1800$ cents, or \$18.00. Questioning leads Mr. Robinson to believe that Olivia's understanding does not go much beyond enabling her to use a shortcut for repeated addition, but he considers this approach a good beginning. He wants to make sure that Olivia recognizes connections between what she is doing and the distributive property (Essential Understanding 2a), seeing, for example, that 24×75 is the same as $(16 + 8) \times 75$, or $(16 \times 75) + (8 \times 75)$.

Using carefully selected computations

A second way of building students' understanding is by questioning them about carefully chosen computations. After creating the problem of the fourth graders who brought 75 cents apiece for a party, Mr. Robinson might develop a series of computations and accompanying questions for his students to reason about to help them build their understanding. Suppose that Mr. Robinson selects the following sequence of computational questions:

- How are 2×7 and 2×70 related?

  Mr. Robinson knows that students need to understand that the second product is not formed by "adding a zero" to the first product, but by becoming "10 times greater" than the first product. The product changes by a factor of 10, from 14 to 14 tens, or 140. Mr. Robinson knows that students can intuitively use the associative property to show how the factor of 10 emerges, and he introduces the formal name of this property after students use it in their reasoning:

  $2 \times 70 = 2 \times (7 \times 10) = (2 \times 7) \times 10 = 14 \times 10$

- How are 2×70 and 2×75 related?

  Although they do not yet know the distributive property, students use it informally as they draw on their understanding of multiplication and number decomposition to show the connection:

  $2 \times 75 = 2 \times (70 + 5) = (2 \times 70) + (2 \times 5)$

- How are 2×70 and 20×70 related?

  Because 20 is ten times 2, 20×70 is ten times 2×70:

  $20 \times 70 = (2 \times 10) \times 70 = (10 \times 2) \times 70 = 10 \times (2 \times 70)$

- How are 2×75 and 20×75 related?

  $20 \times 75 = (2 \times 10) \times 75 = (10 \times 2) \times 75 = 10 \times (2 \times 75)$

- How are 20×75 and 24×75 related?

  $24 \times 75 = (20 + 4) \times 75 = (20 \times 75) + (4 \times 75)$

**Essential ←
Understanding 2a**

The commutative and associative properties of multiplication and the distributive property of multiplication over addition ensure flexibility in computations with whole numbers and provide justifications for sequences of computations with them.

Mr. Robinson makes use of properties and computational fluency to generate questions to help his students to develop their flexibility in computation (Essential Understanding 2a), while simultaneously emphasizing correct mathematical terminology that contributes to mathematical understanding (Essential Understanding 1f).

Working with an area model

A third way of building students' understanding applies their knowledge of area as length times width. If the length of a rectangle is 75 units and the width is 24 units, students can develop a rectangular model to represent the area, such as that shown in figure 3.1. Such an area model provides a geometric representation of the distributive property.

Fig. 3.1. Using an area model to illustrate a procedure for multiplying two-digit numbers

A student who represents her thinking with an area model, which might develop naturally from her use of base-ten materials to explain her thinking, could say something like the following:

> First, I can think about 20 + 4 times 70. I know that 20×70 is 1400 because 20×70 is ten times 2×70. And 4 times 70 is 280 because 4×70 is 10 times 4×7. Then 20×5 is 100, and 4×5 is 20. By doing this in parts and using what I know, I can justify that $1400 + 280 + 100 + 20$ gets the correct answer of 1800.

When students recognize and can use connections among parts of the area model and the symbolic procedures for multiplying, they demonstrate a rich mathematical understanding.

Looking—and Listening— for Understanding

As students extend their knowledge, the teacher's challenge is to provide opportunities to develop more in-depth understandings. Consider Ms. Wojak's approach, as she explains it to Ms. Avery.

Ms. Wojak relates to Ms. Avery an earlier conversation in her classroom. One student, Rena, described her strategy for solving a problem that required sharing 30 cookies among three people. Rena said, "OK, 3 × 10. You give them 10 because you have 30 cookies. If you take 10, 10, and 10, that equals 30. So, there are three people; each can get 10. I switched it around, so it was 30 and 3, and that equaled 10." Ms. Wojak explains that when she asked Rena what she did with the 30, Rena said, "I split 30 into tens. I knew 3 × 10 is 30, so 30 has three tens." Ms. Wojak tells Ms. Avery that at that point, she asked Rena, "What operation symbol would represent that idea of 'splitting into equal parts,' and how would we represent your thinking with an equation?" Rena wasn't sure, so Ms. Wojak posed the following problem:

> If 18 peanut butter cookies are shared among three people, how many cookies will each person get?

Rena said, "Six. I know that 3 × 6 is 18, and if I divide 18 into 3 groups, I will have 6." This time, Rena used the word *divide* instead of *split*, and Ms. Wojak represents her thinking by writing 3 × 6 = 18 and 18 ÷ 3 = 6.

The classroom conversation continued with other students using their understanding of multiplication to derive division sentences. Ms. Wojak explains to Ms. Avery that her students were beginning to develop an understanding of the inverse relationship between multiplication and division (Essential Understanding 1*e*), and she emphasized the terms *multiplication*, *division*, *factor*, *multiple*, and *product* throughout the lesson (Essential Understanding 1*f*). At the end of the discussion, Dashawn used the right distributive property over division to explain his reasoning (Essential Understanding 2*b*), explaining, "Twenty-one divided by 3 is the same as 15 divided by 3 plus 6 divided by 3." Ms. Wojak asked him to write the equation that he had verbalized:

$$21 ÷ 3 = (15 ÷ 3) + (6 ÷ 3)$$

Hallmarks of Teaching with Understanding

Classroom situations inevitably bring attention to a variety of issues in teaching about multiplication and division, including the following:

- The use of "tricks" in computation
- Opportunities for rich mathematical tasks
- The need for mathematical terms and definitions

← **Essential Understanding 1***e*

Division is defined by its inverse relationship with multiplication.

← **Essential Understanding 1***f*

Using proper terminology and understanding the division algorithm provide the basis for understanding how numbers such as the quotient *and* remainder *are used in a division situation.*

← **Essential Understanding 2***b*

The right distributive property of division over addition allows computing flexibly and justifying computations with whole numbers, but there is no left distributive property of division over addition and no commutative or associative property of division of whole numbers.

When teachers see their students' experiences of multiplication and division through their own understanding of the big ideas and essential understandings, they discover ways to turn these three issues into new opportunities for their students to develop their understanding of multiplication and division.

Explaining novel procedures

Students who focus on getting correct answers quickly, without reasoning, come to believe that mathematics requires mostly "tricks" and memorization of procedures. This kind of learning sacrifices understanding. A learning environment that focuses on understanding presents students with no tricks, shortcuts, or mysterious procedures, and the teacher's challenge is to create opportunities for students to use and develop understanding by exploring why "tricks" or memorized procedures make sense.

Consider the following procedure for multiplying by 9. If you number your fingers left to right from 1 through 10 as shown in figure 3.2, and hold finger 2 down, you have one finger to its left and eight fingers to its right. This arrangement "shows" the product of 9 and 2 as 18. If you hold finger 6 down, you have 5 fingers to its left and 4 fingers to its right. This arrangement "shows" that 9×6 is 54.

Fig. 3.2. Representing 9×2 by using a common shortcut that need not be a meaningless procedure for multiplying by 9

What understanding can using this procedure reinforce? What mathematics can we bring to this procedure to make sense of it? One way to reason about and justify the procedure follows:

$9 \times n = (10 - 1) \times n = 10n - 1n$ By using what we know and the distributive property

$9 \times n = 10n - 1n = 10n - 10 + 10 - 1n$ By adding and subtracting 10

$9 \times n = 10n - 10 + 10 - 1n = 10(n - 1) + (10 - n)$ By using the distributive property again

In the hand procedure for multiplying $9 \times n$—

  • n is the number of the finger held down;

- $n - 1$ represents the number of fingers to the left of this finger;

- $10 - n$ represents the number of fingers to its right; and

- $10(n - 1) + (10 - n)$ is equal to $9 \times n$.

Thus, our use of the distributive property provides a mathematical explanation for this "trick" with fingers (Essential Understanding 2*a*).

Teachers who teach for conceptual understanding might explore with their students why tricks, shortcuts, and seemingly mysterious procedures work. It is through such exploration that students delve into mathematics and remember it in the long run. It is important to note that the reasoning processes behind the tricks, shortcuts, and procedures may take time to develop. However, these reasoning processes can reinforce students' understanding of mathematics encountered outside the classroom and within previous mathematics lessons. In fact, students can have quite a bit of fun determining how they can use the mathematics that they understand to explain "tricks" to people who won't take time to think!

Using worthwhile tasks

Teachers can create tasks to motivate students to reason, promoting their understanding in many ways. For example, a fifth-grade teacher, Mr. Charles, puts up an interactive bulletin board covering the entire back wall of his classroom. The "four 4s" board displays the numerals representing the numbers from 1 to 100. The students' objective is to find as many expressions as they can for each number. They can use any operations but only four 4s. Some examples for 1 and 60 follow:

$$\frac{4 \times 4}{4 \times 4} \qquad \frac{44}{44} \qquad 4 \times 4 \times 4 - 4 \qquad 44 + 4 \times 4$$

Several expressions are possible for each number. Mr. Charles's students discover that order of operations, number relationships involving multiplication and division, and the properties of operations all play roles in finding alternative ways of expressing the same number. Periodically, Mr. Charles discusses these ideas to help his students understand the properties and relationships.

Taking advantage of a different setting, Carey (1992) suggests using children's literature to motivate students to solve problems. Using the book *The Patchwork Quilt* (Flournoy 1985), Carey proposes tasks involving multiplication and division concepts. By carefully wording problems, she encourages students to use multiple representations for multiplication and division situations. For example, an array representation can model one way of thinking about the problem, "How many squares will Tanya need if she makes a quilt that has 6 rows with 4 squares in each row?" (Carey 1992, p. 200).

← Essential Understanding 2*a*

The commutative and associative properties of multiplication and the distributive property of multiplication over addition ensure flexibility in computations with whole numbers and provide justifications for sequences of computations with them.

By thoughtfully selecting the numbers for each problem, Carey encourages thinking about factors, commutativity, and concepts involving the relationship of area and perimeter. An extension task on scale drawing makes connections to the concept of scalar multiplication. If the length and width of a rectangle are both doubled, as shown in figure 3.3, what is the effect on the perimeter of the shape and the area of the rectangular space? Thiessen (1998, 2004) details additional connections to literature that incorporate concepts related to multiplication and division.

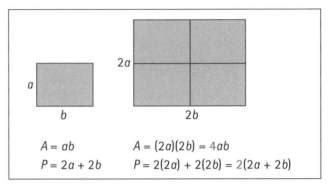

Fig. 3.3. Diagrams and expressions illustrating the effects of doubling the length and width of a rectangle on its area and perimeter

Topics from units that students are working on in science, social studies, and art can also provide contexts for reasoning about multiplication and division, as can topics related to students' school life in general. For example, consider the following situation:

> If 35 third-grade students bring $25 each to purchase a school jacket, how much money is that? If 45 fourth-grade students bring $25 each for the same purpose, how many more dollars do the fourth-graders bring for jackets?

Students should be able to answer the second question with minimal calculations and no paper and pencil.

Also consider presenting problems to students that connect to their interests outside school. The following is just one example:

> For a Super Bowl, _____ rings were purchased by the winning team. If each ring costs $ _____ , how much was the bill for all of the rings?

Information that students might use for the Super Bowl problem can be found at http://www.nola.com/superbowl/index.ssf/2010/06/new_orleans_saints_super_bowl_11.html.

Students could research the situation for homework and bring in actual data so that the problem to be solved reflects the current real-life situation. Strategy sharing can also be a mental mathematics activity. In Mr. Lee's class, the students found that 150 rings were purchased at a cost of $5,000 each. One student, David, reasoned that 100 rings would cost 100 × 5,000, or $500,000. Fifty more rings would cost half as much, or $250,000. So David came up with a total cost of $750,000.

Meanwhile, another student, Marianne, began by considering that 5×15 is 75. She knew that the number of dollars needed would be ten thousand times greater than 75. So she reasoned that the total cost would be $750,000. Her calculation can be justified by using properties of whole number multiplication and addition:

$$10,000 \times (5 \times 15) = 1,000 \times (5 \times 15) \times 10 = (1,000 \times 5) \times (15 \times 10) = 5,000 \times 150$$

When students share reasoning strategies on a daily basis, they have opportunities to build on their own strategies. For all students, an important part of the sharing of strategies is questioning and explaining as they develop a deeper and richer understanding.

Using appropriate language and definitions

As students progress from grade 3 to 4 to 5 and continue to discuss strategies, teachers should use the same terms to ensure effective communication about the mathematics that the students are learning. Confusion among students can result if one teacher says *turn-arounds* and another says *commutative*. If a teacher does not use a mathematical term precisely, students may use the concept incorrectly. For example, they may apply the commutative property of multiplication to the operation of division—an error that teachers often overhear students making in discussions about division.

Teachers can help students develop communication and reasoning skills by referring to terms and definitions that the students are familiar with but have neglected to use as part of their shared reasoning process. Consider the following scenario. Kobe is confused, so his teacher, Mr. Dean, engages him in a discussion:

Kobe: I can't remember if $5 \div 0 = 0$, or $0 \div 5 = 0$. Or are they both 0?

Mr. Dean: What do we mean when we say that we *divide*?

Kobe: [*Thinking*] Let's see... If $12 \div 3 = 4$, then $12 = 3 \times 4$... So, for $5 \div 0 = 0$, then $5 = 0 \times 0$. That doesn't make sense. Let's try $0 \div 5 = 0$, so $0 = 5 \times 0$. That makes sense. So $0 \div 5 = 0$ makes sense, but $5 \div 0 = 0$ does not.

Mr. Dean: [*Summarizing*] We call $5 \div 0$ *undefined* because division by zero does not make sense when we use the definition of division.

In this scenario, the teacher helped the student by using an idea that the class had studied to help make sense of division involving zero. The teacher also introduced the term *undefined* so that the student could extend his mathematical vocabulary. Mathematical reasoning and discourse are enhanced when teachers use correct terms and help students use definitions that their previous work has helped them to understand.

Making Time to Foster Understanding

Developing understanding requires not only worthwhile tasks and appropriate language but also time. Consider the time that Mr. Fuentes devotes to helping his students develop an understanding of division computation. He presents the following problem to his class:

> Mr. Fuentes has 698 pencils. He is going to give each of the fifth graders a box of 12 pencils. How many boxes will Mr. Fuentes be able to fill?

Mr. Fuentes makes sure all of the students understand the problem and encourages them to get any materials that they need to work on it. Then he walks around the room observing students' work. When Mr. Fuentes approaches Alyssa, he sees that she has written the expressions that appear in figure 3.4.

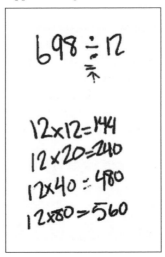

Fig. 3.4. Alyssa's initial work on Mr. Fuentes's problem

Her written work makes Mr. Fuentes think that Alyssa understands that division has an inverse relationship with multiplication (Essential Understanding 1e) and recognizes that multiplying by tens is efficient. He notices her calculation error, but without calling attention to it, he says, "I see what you've done. Explain your reasoning to me. What number are you trying to get to?" Alyssa explains that she is trying to get to 698 and that she needs to figure out how many multiples of 12 are in that number. She knows that 10×12 is 120, and she uses that product to derive 12×20 and doubles her new product to get 12×40. She says, "And that's still not high enough, so I'm going to do 12×80." She miscalculates again and says, "That's still not high enough." Mr. Fuentes says,

"Wait a minute, if 12 × 40 is 480, then how did you get 12 × 80 is 560?" Alyssa thinks and says, "That's wrong, I forgot to double the 4... That would be too much."

At this point, Mr. Fuentes wants to use Alyssa's thinking, in addition to her knowledge of both basic facts and multiplying by 10s and 100s, to help her to develop a more efficient algorithm. He begins a dialogue with Alyssa to help her make connections from the mathematics that she knows to the mathematics he wants her to know. He uses a series of questions as he writes out significant numbers from her thinking process:

Mr. Fuentes: You want to know how many 12s are in 698 and you want to get there quickly.

Alyssa: [*Nods.*]

Mr. Fuentes: And you know that for every ten 12s, it's 120, so let's work with this. How many 12s are in 698, just looking at it? How many times are you going to get 120 out of there?

Alyssa: [*Estimating*] Four.

Mr. Fuentes: But you already did 40 times 12 and didn't get to 698.

Alyssa: [*Estimating*] Maybe 50.

Mr. Fuentes: So, if you do 50 × 12, what will that get you?

Alyssa: [*Hesitating.*]

Mr. Fuentes: [*Waits before asking another question*] What do you know about 5 × 12?

Alyssa: It's 60.

Mr. Fuentes: So does that help with 50 × 12?

Alyssa: I think so. It's times 10 more, so that would be 600!

Mr. Fuentes: I love that "times 10"—you're not just going to add a zero! So now, look, you're almost there. What do you know at this point?

Alyssa: I only have 98 more to use.

Mr. Fuentes: Yes, so what's a good estimate?

Alyssa: Five 12s are 60. So I would have 38 more to get to 98.

Mr. Fuentes: So, what do you know about 38?

Alyssa: Well, 3 × 12 is 36.

Mr. Fuentes: So, what do you know about this problem for an answer to how many boxes Mr. Fuentes can fill?

Alyssa: It's 58 boxes.

Mr. Fuentes: Why?

Alyssa: Because 50 × 12 is 600, then I did 5 × 12 is 60,
                 and then I did 3 × 12, which is 36.

Mr. Fuentes: Will all the pencils be in a box?

Alyssa: No, there would be 2 left.

Mr. Fuentes: So, if Mr. Fuentes has to have all the pencils in a
                 box, how many boxes will he need?

Alyssa: Um, 59.

Mr. Fuentes: Excellent.

At this point, Mr. Fuentes gives Alyssa another problem to assess whether she is beginning to develop a more efficient strategy:

> If we have 8,492 pencils and we put 12 in a box, how many
> boxes exactly will we need?

Alyssa begins writing; her work appears in figure 3.5.

Fig. 3.5. Calculations involved in Alyssa's solution to the second problem that Mr. Fuentes poses

Mr. Fuentes can see that after Alyssa has written 8,492, she
begins to subtract groups of 12. She seems to remember that 5 ×
12 is 60 and then reasons that 500 × 12 is 60 hundreds, or 6,000.
Alyssa subtracts 6,000 from 8,492 to get 2,492, writing 500 above
to keep track of the number of groups of 12 in her quotient. She
realizes that she is left with 2,492 and stops. Mr. Fuentes begins his
questioning:

Mr. Fuentes: [*Pointing to the divisor of 12 and the difference of
                 2492*] Twelve and 2,492—I see a number that should
                 pop right into your head; what is it?

Alyssa: Twenty-four popped into my head because it is the
                 double of 12, so 20 pops into my head, 20 × 12 is

240, and 200 × 12 is 2,400. I have 92 left. OK, and I know that 10 × 12 is 120, and that's too much. And 9 × 12 is 108. But I can take away 12 to get 96, and then take away 12 to get 84, so 7 × 12 is 84.

Mr. Fuentes: How are you going to use that idea?

Alyssa: I'll take that away from the 92 [*writes 84 under 92 on the board and continues writing*], and 92 – 84 = 8, so that would be 8 left.

Mr. Fuentes: So what do you conclude as the answer to the number of boxes that we will need?

Alyssa: [*Adding the partial quotients and apparently putting the remaining 8 pencils in a separate box*] It's 708!

Mr. Fuentes understands that knowledge of the multiplication facts connected to 12 is helpful to his students' development of this efficient strategy. Furthermore, Mr. Fuentes realizes that his students are developing a fuller understanding of division as they think about a more efficient strategy. He also knows that helping students acquire this rich understanding takes time.

Embedding Assessment in Instruction

To assess a student's understanding of multiplication and division, a classroom teacher needs to consider each observation, discussion, end-of-unit assessment, work sample, and problem-solving session thoughtfully and critically. Teaching and assessing for understanding are inextricably related. Teachers need to probe students' understanding consistently as they observe them and engage them in dialogue. Through observation, effective teachers can gain critical information about their students' understanding.

Consider how a third-grade teacher, Mr. Henry, uses questioning as a means to assess a student's understanding of division. Mr. Henry gives his students the following problem:

If 3 friends share 18 cookies, how many cookies does each get?

Mr. Henry notes that one student, Tierek, who has begun representing the situation with cubes, uses familiar facts and relationships to get close to the answer. Then he adjusts to get the exact answer.

Because Mr. Henry is interested in using his students' level of understanding as a means of informing instruction, he questions Tierek, probing his reasoning until he is confident that he has identified the depth of Tierek's understanding. Notice how Mr. Henry bases his questions on what Tierek does and says in solving the problem:

Mr. Henry: I noticed that first you grouped the cubes in 5s. Why

Big Idea 1

Multiplication is one of two fundamental operations, along with addition, which can be defined so that it is an appropriate choice for representing and solving problems in many different situations.

➡ **Essential Understanding 1c**
A situation that can be represented by multiplication has an element that represents the scalar and an element that represents the quantity to which the scalar applies.

➡ **Essential Understanding 1e**
Division is defined by its inverse relationship with multiplication.

➡ **Essential Understanding 2a**
The commutative and associative properties of multiplication and the distributive property of multiplication over addition ensure flexibility in computations with whole numbers and provide justifications for sequences of computations with them.

did you do that and then stop after you'd built three groups?

Tierek: Because I know how to count by 5s, I thought I'd grab them in 5s.

Mr. Henry: You also know that three groups of 5 is 15, so why did you check how many cubes you had out for the cookies by counting 5, 10, 15?

Tierek: Well, I guess I just had to double check.

Mr. Henry: I see. After that, you took out only three more unit cubes and added one to each pile. Why did you stop after that?

Tierek: Because I knew that I had only 15 cookies there going to 3 kids—the piles are the kids—and there are supposed to be 18 cookies and 3 kids. So if I added 3 more it would make the eighteen. The answer is 6.

Mr. Henry poses his questions to gain understanding of Tierek's thinking. Knowledge of a student's thinking, together with a deep understanding of multiplication and division, enables a teacher to create learning tasks that provide opportunities for the student to make sense of representations and reasoning strategies.

Jotting down key instances of students' thinking, their use of mathematical vocabulary, and their work with models or symbolic procedures for representing multiplication and division provides an invaluable written account of how a student's understanding develops over time. Perhaps Tierek has never before used groups of 5—something he knows well—to get close to the answer so he can adjust his work to get the exact answer. His thinking is an example of how students make informal use of the distributive property as they begin to understand the mathematics more formally.

As Mr. Henry listens and draws on his understanding of the big ideas and essential understandings of multiplication and division, he can ask a variety of questions—each based on an essential understanding that he has developed—to assess his students' understanding, including the following:

- Do they understand when to multiply or divide? (Big Idea 1)

- Do they understand how to represent various situations involving multiplication and division? (Essential Understanding 1c)

- Do they understand that multiplication can be used to help them solve division problems? (Essential Understanding 1e)

- Do they understand the properties of multiplication and know when to use them efficiently? (Essential Understanding 2a)

The answers to these questions can help Mr. Henry decide if his students need additional experience solving routine word problems with a variety of structures. These problems can help his students gain a better understanding of when to divide and give them opportunities to use multiplication to help solve division problems.

Conclusion

The task of teaching multiplication and division concepts is a challenging one. Because these concepts provide a foundation for mathematics far beyond grades 3–5, it is critical that students have opportunities to develop the necessary understanding for future mathematics learning.

This chapter presents several aspects of effective instruction that can help students develop such a foundation. It is important that teachers understand both a variety of strategies as well as how students' thinking can be developed to levels that reflect a flexible and growing understanding. Teachers' knowledge of mathematics is critical to this process because their deep understanding can serve as a conduit through which students develop their own robust understanding.

References

Barnett-Clarke, Carne, William Fisher, Rick Marks, and Sharon Ross. *Developing Essential Understanding of Rational Numbers for Teaching Mathematics in Grades 3–5*. Essential Understanding Series. Reston, Va.: National Council of Teachers of Mathematics, 2010.

Bates, Tom, and Leo Rousseau. "Will the Real Division Algorithm Please Stand Up?" *Arithmetic Teacher* 33 (March 1986): 42–46.

Blanton, Maria, Linda Levi, Terry Wayne Crites, and Barbara J. Dougherty. *Developing Essential Understanding of Algebraic Thinking for Teaching Mathematics in Grades 3–5*. Essential Understanding Series. Reston, Va.: National Council of Teachers of Mathematics, 2011.

Caldwell, Janet H., Karen Karp, and Jennifer M. Bay-Williams. *Developing Essential Understanding of Addition and Subtraction for Teaching Mathematics in Prekindergarten–Grade 2*. Essential Understanding Series. Reston, Va.: National Council of Teachers of Mathematics, 2011.

Caliandro, Christine Koller. "Children's Inventions for Multidigit Multiplication and Division." *Teaching Children Mathematics* 6 (February 2000): 420–24, 426.

Carey, Deborah A. "*The Patchwork Quilt:* A Context for Problem Solving." *Arithmetic Teacher* 40 (December 1992): 199–203.

Cooney, Thomas J., Sybilla Beckmann, and Gwendolyn M. Lloyd. *Developing Essential Understanding of Functions for Teaching Mathematics in Grades 9–12*. Essential Understanding Series. Reston, Va.: National Council of Teachers of Mathematics, 2010.

Dougherty, Barbara J., Alfinio Flores, Everett Louis, and Catherine Sophian. *Developing Essential Understanding of Number and Numeration for Teaching Mathematics in Prekindergarten–Grade 2*. Essential Understanding Series. Reston, Va.: National Council of Teachers of Mathematics, 2010.

Englert, Gail R., and Rose Sinicrope. "Making Connections with Two-Digit Multiplication." *Arithmetic Teacher* 41 (April 1994): 446–48.

Flournoy, Valerie E. *The Patchwork Quilt*. New York: Dial Books for Young Readers, 1985.

Fuson, Karen C. "Toward Computational Fluency in Multidigit Multiplication and Division." *Teaching Children Mathematics* 9 (February 2003): 300–305.

Graeber, Anna O. "Research into Practice: Misconceptions about Multiplication and Division." *Arithmetic Teacher* 40 (March 1993): 408–11.

Hendrickson, A. Dean. "Verbal Multiplication and Division Problems: Some Difficulties and Some Solutions." *Arithmetic Teacher* 33 (April 1986): 26–33.

Lannin, John, Amy B. Ellis, and Rebekah Elliott. *Developing Essential Understanding of Mathematical Reasoning for Teaching Mathematics in Prekindergarten–Grade 8*. Essential Understanding Series. Reston, Va.: National Council of Teachers of Mathematics, forthcoming.

Lobato, Joanne, and Amy B. Ellis. *Developing Essential Understanding of Ratios, Proportions, and Proportional Reasoning for Teaching Mathematics in Grades 6–8*. Essential Understanding Series. Reston, Va.: National Council of Teachers of Mathematics, 2010.

National Council of Teachers of Mathematics (NCTM). *Principles and Standards for School Mathematics*. Reston, Va.: NCTM, 2000.

———. *Curriculum Focal Points for Prekindergarten through Grade 8 Mathematics: A Quest for Coherence*. Reston, Va.: NCTM, 2006.

———. *Focus in High School Mathematics: Reasoning and Sense Making*. Reston, Va.: NCTM, 2009.

National Mathematics Advisory Panel. *Foundations for Success: The Final Report of the National Mathematics Advisory Panel*. Washington, D.C.: U.S. Department of Education, 2008.

Quintero, Ana Helvia. "Children's Conceptual Understanding of Situations Involving Multiplication." *Arithmetic Teacher* 33 (January 1986): 34–37.

Sowder, Judith T. "Estimation and Number Sense." In *Handbook of Research on Mathematics Teaching and Learning*, edited by Douglas A. Grouws, pp. 371–89. New York: Macmillan, 1992.

Thiessen, Diane. *The Wonderful World of Mathematics: A Critically Annotated List of Children's Books in Mathematics*. 2nd ed. Reston, Va.: National Council of Teachers of Mathematics, 1998.

Thiessen, Diane. *Exploring Mathematics through Literature: Articles and Lessons for Prekindergarten through Grade 8*. Reston, Va.: National Council of Teachers of Mathematics, 2004.

Weiland, Linnea. "Matching Instruction to Children's Thinking about Division." *Arithmetic Teacher* 33 (December 1985): 34–35.